I0473858

Table of Contents

GROWING EDIBLE MUSHROOMS

Introduction

Mushrooms are a group of fleshy saprophytic fungi that are found growing on dead organic matter. Over 10,000 species of mushrooms are believed to be found naturally growing in moist, damp forests and soils rich in organic matter throughout the world. Among these, some are edible while others are toxic. A few species of mushrooms are known for their medicinal properties as well and hence they are known as medicinal mushrooms.

A growing mushroom initially forms a minute fruiting body during its initial developmental stages. This fruiting body is called a pin because of its small size and this developmental stage is known as pin-stage. Later when these pins grow in size, they are called buttons. This stage during mushroom development is known as button-stage. Once buttons are formed, mushrooms grow or expand rapidly into enlarged fruiting bodies and caps. Some mushrooms expand overnight while others grow slowly.

Edible Mushrooms

Edible mushrooms are highly nutritious, high in essential amino acids and are with absolutely nil or less fat content. There is no cholesterol in them; they are easily digestible due to the presence of high fiber content, and can be compared with meat in their nutritional value. Edible mushrooms are often grouped along with vegetables and known as 'Meat of the Vegetable World'.

Edible mushrooms are easy to cultivate. While cultivating edible mushrooms, a number of substrate materials may be used as their food materials such as paddy straw, saw dust, cotton waste, cocoa bean shell, and wooden logs. Most of the edible mushrooms like milky mushrooms, shiitake, Enoki, oyster mushrooms, paddy straw mushrooms, shaggy manes,

Maitake mushrooms etc are members of the order Agaricales. They are suited for home-growing as well as for commercial production. A list of the major edible mushrooms is given below:

1   Milky Mushrooms (Calocybe indica)

2   Paddy Straw Mushrooms (Volvariella volvacea)

3   White Button Mushrooms (Agaricus bisporus)

4   The Brown Mushrooms (Agaricus bisporous Brown strain)

5   Shiitake Mushrooms (Lentinula edodes)

6   Oyster mushrooms (Pleurotus spp)

7   Enoki Mushrooms (Flammulina velutipes)

8   Maitake Mushrooms (Grifola frondosa)

9   Morel Mushrooms (Morchella esculenta)

10 Chanterelle Mushrooms (Cantharellus cibarius)

Milky Mushrooms (Calocybe indica)

Milky mushrooms are tropical in habit and are very popular in India. It is found growing naturally in the damp, moist forest environments of Indian subcontinent. Favorable temperature range for its growth is considered to be 25-30° C. Nutritional value of 100 grams of raw Milky mushrooms is given below:

Nutrient values and weights are for 100 grams of edible portion	
Carbohydrates	10.21 g
Crude fiber	1.12 g
Dietary fiber (insoluble)	41.05 g
Dietary fiber (soluble)	0.74 g

Energy	51.46 kcal
Fat	0.15g
Moisture	84.84 g
Protein	2.31g

Paddy Straw Mushrooms (Volvariella volvacea)

Paddy Straw mushrooms are subtropical in habit and are successfully grown in areas with high rainfall. Favorable temperature range for its growth is considered to be 30 - 35°C. They are very popular in India, China, Taiwan, Thailand and other South East Asian countries. Substrate material popularly used for its cultivation is paddy straw and therefore the name 'paddy straw mushrooms'.

White Button Mushrooms (Agaricus bisporus)

White button mushrooms are one of the most widely consumed mushrooms in the world. They are believed to be cultivated in more than 100 countries and are known by the common names such as button mushrooms, white mushrooms, and table mushrooms. Good compost and good quality spawns and a simulated environment are absolute necessities for successful growing of White button mushrooms. Favorable temperature range for its growth is considered to be $24°$ C and 15 -17° C. Nutritional value of 100 grams of raw White Button mushrooms is given below:

Nutrient values and weights are for edible portion		
Nutrient	Unit	Value per 100.0g
Proximates		
Water	g	92.45
Energy	kcal	22
Protein	g	3.09
Total lipid (fat)	g	0.34

Carbohydrate, by difference	g	3.26
Fiber, total dietary	g	1
Sugars, total	g	1.98
Minerals		
Calcium, Ca	mg	3
Iron, Fe	mg	0.5
Magnesium, Mg	mg	9
Phosphorus, P	mg	86
Potassium, K	mg	318
Sodium, Na	mg	5
Zinc, Zn	mg	0.52
Vitamins		
Vitamin C, total ascorbic acid	mg	2.1
Thiamin	mg	0.081
Riboflavin	mg	0.402
Niacin	mg	3.607
Vitamin B-6	mg	0.104
Folate DFE a	Mcg_DFE	17
Vitamin B-12	µg	0.04
Vitamin E (alpha-tocopherol)	mg	0.01
Vitamin D (D2 + D3)	µg	0.2
Vitamin D	IU	7
Lipids		
Fatty acids, total saturated	g	0.05
Fatty acids, total monounsaturated	g	0
Fatty acids, total polyunsaturated	g	0.16
Cholesterol	mg	0
Ergosterol	mg	56

Source: USDA Nutrient Database

The Brown Mushrooms (Agaricus bisporous Brown strain)

Crimini and Portabella mushrooms are known as the brown mushrooms. Crimini mushrooms are also known as Swiss brown mushroom, Roman brown mushroom, Italian brown and Italian mushrooms. They are believed to be originated in Europe and North America. Crimini mushrooms have a darker or brown color, deeper flavor and denser texture than the button mushrooms. Nutritional value of 100 grams of edible raw Crimini mushrooms is given below:

Proximates	
Water(g)	*92.12*
Energy(kcal)	*22*
Protein(g)	*2.5*
Total lipid (fat)(g)	*0.1*
Carbohydrate, by difference(g)	*4.3*
Fiber, total dietary(g)	*0.6*
Sugars, total(g)	*1.72*
Minerals	
Calcium, Ca (mg)	*18*
Iron, Fe(mg)	*0.4*
Magnesium, Mg(mg)	*9*
Phosphorus, P(mg)	*120*
Potassium, K(mg)	*448*
Sodium, Na (mg)	*6*
Zinc, Zn (mg)	*1.1*
Vitamins	
Thiamin (mg)	*0.095*
Riboflavin (mg)	*0.49*
Niacin (mg)	*3.8*

Vitamin B-6 (mg)	*0.11*
Folate, DFE a (mcg_DFE)	*25*
Vitamin B-12 (µg)	*0.1*
Vitamin E (alpha-tocopherol)(mg)	*0.01*
Vitamin D (D2 + D3) (µg)	*0.1*
Vitamin D (IU)	*3*
Lipids	
Fatty acids, total saturated (g)	*0.014*
Fatty acids, total monounsaturated(g)	*0.002*
Fatty acids, total polyunsaturated(g)	*0.042*
Cholesterol (mg)	*0*
Ergosterol = 62 mg/100 g.	

Source: USDA Nutrient Database

Portabella mushrooms are mature Crimini mushrooms, which are three to seven days older than the Crimini. Because of their extended growing period, portabellas have larger caps and deeper color. Ideal temperature for the growth of both the Crimini and the portabellas are considered to be 24°C and 15-17°C. Nutritional value of 100 grams of raw Portabella mushrooms is given below:

Proximates	
Water(g)	*92.82*
Energy (kcal)	*22*
Protein (g)	*2.11*
Total lipid (fat)(g)	*0.35*
Carbohydrate, by difference (g)	*3.87*
Fiber, total dietary (g)	*1.3*
Sugars, total (g)	*2.5*
Minerals	

Calcium, Ca (mg)	3
Iron, Fe(mg)	0.31
Phosphorus, P(mg)	108
Potassium, K(mg)	364
Sodium, Na(mg)	9
Zinc, Zn(mg)	0.53
Vitamins	
Thiamin (mg)	0.059
Riboflavin (mg)	0.13
Niacin (mg)	4.494
Vitamin B-6 (mg)	0.148
Folate, DFE a (Mcg_DFE)	28
Vitamin B-12 (μg)	0.05
Vitamin E (alpha-tocopherol) (mg)	0.02
Vitamin D (D2 + D3) (μg)	0.3
Vitamin D (IU)	10
Lipids	
Fatty acids, total saturated (g)	0.06
Fatty acids, total monounsaturated (g)	0.02
Fatty acids, total polyunsaturated (g)	0.117
Cholesterol (mg)	0
Ergosterol = 56 mg/100 g.	

Source: USDA Nutrient Database

Shiitake Mushrooms (Lentinula edodes)

Shiitake Mushrooms are fleshy, brown-colored mushrooms that are known for their medicinal properties. Ideal temperature for the growth of shiitake mushrooms is considered to be 18-24°C. Shiitake mushrooms are believed to be originated in East Asia, particularly in the regions comprising of Japan, Korea and China. These mushrooms are known for both their edible as well as medicinal properties. Shiitake mushrooms are considered to be a good remedy for premature ageing and respiratory problems. The Japanese cultivates shiitake mushrooms on the wooden logs of shii trees. These mushrooms can be consumed both as fresh and dried forms. Dried shiitakes can be rehydrated by soaking them in water for a few minutes. Sun dried shiitake mushrooms are in more demand as drying process breaks down proteins into amino acids and converts ergosterol present in them to vitamin D. Nutritional value of 100 grams of raw Shiitake mushrooms is given below:

Proximates		
Water (g)	*89.74*	
Energy(kcal)	*34*	
Protein(g)	*2.24*	
Total lipid (fat) (g)	*0.49*	
Carbohydrate, by difference(g)	*6.79*	
Fiber, total dietary(g)		*2.5*
Sugars, total(g)	*2.38*	
Minerals		
Calcium, Ca (mg)	*2*	
Iron, Fe(mg)	*0.41*	
Magnesium, Mg(mg)		*20*
Phosphorus, P(mg)	*112*	
Potassium, K(mg)	*304*	
Sodium, Na(mg)	*9*	
Zinc, Zn(mg)	*1.03*	
Vitamins		
Thiamin(mg)	*0.015*	
Riboflavin(mg)	*0.217*	
Niacin(mg)	*3.877*	

Vitamin B-6(mg)	*0.293*
Vitamin D (D2 + D3) μg	*0.4*
Vitamin D(IU)	*18*
Ergosterol = 85 mg/100 g.	

Source: USDA Nutrient Database

Oyster mushrooms (Pleurotus spp)

Oyster mushrooms can be grown on many substrates, and are very easy to cultivate even for a beginner; it takes hardly one or two months to complete one cycle of production, starting from spawning to harvesting to marketing. Ideal temperature for the growth of oyster mushrooms is considered to be 20-30oC. There is a large type of oyster mushrooms called 'King Oyster Mushrooms'. King oyster mushrooms are very popular in India and known by the name 'Kabul dhingri'. Ideal temperature for the growth of king oyster mushrooms is considered to be 17-20°C. Nutritional value of 100 grams of raw Oyster mushrooms is given below:

Proximates		
Water(g)	*89.18*	
Energy(kcal)	*33*	
Protein(g)	*3.31*	
Total lipid (fat) (g)	*0.41*	
Carbohydrate, by difference(g)	*6.09*	
Fiber, total dietary(g)		*2.3*
Sugars, total(g)	*1.11*	
Minerals		
Calcium, Ca(mg)	*3*	
Iron, Fe(mg)	*1.33*	
Magnesium, Mg(mg)	*18*	
Phosphorus, P(mg)	*120*	
Potassium, K(mg)	*420*	

10

Sodium, Na(mg)	*18*	
Zinc, Zn(mg)	*0.77*	
Vitamins		
Vitamin C, total ascorbic acid(mg)		*0*
Thiamin(mg)	*0.125*	
Riboflavin (mg)	*0.349*	
Niacin(mg)	*4.956*	
Vitamin B-6(mg)	*0.11*	
Folate, DFE a	*mcg_DFE 38*	
Vitamin A, RAE	*mcg_RAE 2*	
Vitamin A, IU	*48*	
Vitamin D (D2 + D3) (μg)		*0.7*
Vitamin D (IU)	*29*	
Lipids		
Fatty acids, total saturated (g)		*0.062*
Fatty acids, total monounsaturated(g)		*0.031*
Fatty acids, total polyunsaturated(g)		*0.123*
Cholesterol(mg)	*0*	
Ergosterol = 64 mg/100 g.		

Enoki Mushrooms (Flammulina velutipes)

Enoki mushrooms or Enokitake are white colored, thin long mushrooms that are popular South East Asian countries such as China, Japan and Korea. They are also known as golden needle mushrooms. Enoki mushrooms are consumed both as fresh and as canned mushrooms. In China, enokitake mushrooms are found naturally growing on the wooden logs of the Chinese Hackberry trees whereas in Japan, they are –known as Enoki mushrooms locally- found growing on the wooden logs of mulberry trees and persimmon trees. Enoki mushrooms can be easily cultivated using saw dust or corn cobs as substrate materials and plastic bottles s containers. A high CO2 environment is required for its cultivation for the production of high-quality long thin mushroom-stems. Nutritional value of 100 grams of raw Enoki mushrooms is given below:

Proximates		
Water(g) 88.34		
Energy(kcal)	37	
Protein(g) g	2.66	
Total lipid (fat) (g)	0.29	
Carbohydrate, by difference(g)	7.81	
Fiber, total dietary(g)		2.7
Sugars, total(g)	0.22	
Minerals		
Iron, Fe(mg)	1.15	
Magnesium, Mg(mg)	16	
Phosphorus, P(mg)	105	
Potassium, K(mg)	359	
Sodium, Na(mg)	3	
Zinc, Zn(mg)	0.65	
Vitamins		
Thiamin(mg)	0.225	
Riboflavin (mg)	0.2	
Niacin(mg)	7.032	
Vitamin B-6(mg)	0.1	
Folate, DFE a(mcg_DFE)		48
Vitamin E (alpha-tocopherol) (mg) 0.01		
Vitamin D (D2 + D3) μg		0.1
Vitamin D IU	5	
Lipids		

Fatty acids, total saturated(g)	*0.02*
Fatty acids, total monounsaturated(g)	*0*
Fatty acids, total polyunsaturated(g)	*0.09*
Cholesterol(mg)	*0*
Ergosterol = 36 mg/100 g.	

Source: USDA Nutrient Database

Maitake Mushrooms (Grifola frondosa)

Maitake mushrooms are believed to be originated in Japan and North America and they are still very popular in Japan. In Japanese language, the word 'Maitake' means "dancing mushroom", a name given due to its growing characteristics; they are perennial fungi and often grow in the same place in succession and occur in clusters at the base of the trees. Common names of Maitake mushrooms are Hen-of-the-Woods, Ram's Head and Sheep's Head. Nutritional value of 100 grams of raw Maitake mushrooms is given below:

Proximates	
Water(g)	*90.37*
Energy(kcal)	*31*
Protein (g)	*1.94*
Total lipid (fat) (g)	*0.19*
Carbohydrate, by difference(g)	*6.91*
Fiber, total dietary(g)	*2.7*
Sugars, total(g)	*2.07*
Minerals	
Calcium, Ca(mg)	*1*
Iron, Fe(mg)	*0.3*
Magnesium, Mg(mg)	*10*
Phosphorus, P(mg)	*74*
Potassium, K(mg)	*204*

Sodium, Na(mg)	1
Zinc, Zn(mg)	0.75
Vitamins	
Thiamin(mg)	0.146
Riboflavin (mg)	0.242
Niacin(mg)	6.585
Vitamin B-6(mg)	0.056
Folate, DFE a (mcg_DFE)	21
Vitamin E (alpha-tocopherol) (mg)	0.01
Vitamin D (D2 + D3) μg	28.1
Vitamin D IU	1123
Lipids	
Fatty acids, total saturated (g)	0.03
Fatty acids, total monounsaturated(g)	0.03
Fatty acids, total polyunsaturated(g)	0.09
Cholesterol(mg)	0
Ergosterol = 72 mg/100 g.	

Source: USDA Nutrient Database

Morel Mushrooms (Morchella esculenta)

Morels are one of the most delicious mushrooms varieties in the world and are highly popular in France. They are used extensively in many types of French cuisines. They are commonly known by their popular name 'sponge mushrooms' and are very popular in Germany and USA also. Nutritional value of 100 grams of raw Morel mushrooms is given below:

Proximates	
Water(g)	89.61
Energy(kcal)	31

Protein(g)	3.12	
Total lipid (fat) (g)	0.57	
Carbohydrate, by difference(g)	5.1	
Fiber, total dietary(g)	2.8	
Sugars, total(g)	0.6	
Minerals		
Calcium, Ca(mg)	43	
Iron, Fe(mg)	12.18	
Magnesium, Mg(mg)	19	
Phosphorus, P(mg)	194	
Potassium, K(mg)	411	
Sodium, Na(mg)	21	
Zinc, Zn(mg)	2.03	
Vitamins		
Thiamin(mg)	0.069	
Riboflavin (mg)	0.205	
Niacin(mg)	2.252	
Vitamin B-6(mg)	0.136	
Vitamin D (D2 + D3) µg	5.1	
Vitamin D (IU)	206	
Lipids		
Fatty acids, total saturated (g)	0.065	
Fatty acids, total monounsaturated(g)		0.052
Fatty acids, total polyunsaturated(g)		0.433
Ergosterol = 26 mg/100 g		
Brassicasterol = 29 mg/100 g.		

Source: USDA Nutrient Database

Chanterelle Mushrooms (Cantharellus cibarius)

Chanterelle mushrooms are a group of golden-colored, fleshy, funnel-shaped edible fungi known for their excellent edible properties. They are very popular in Europe, North America, Africa and Asia. They are found naturally occurring in moist, temperate coniferous forests. Chanterelle mushrooms are normally consumed after cooking and they are suitable for drying also. Dried chanterelle mushrooms are powdered and used for seasoning soups and sauces. Nutritional value of 100 grams of raw Chanterelle mushrooms is given below:

Proximates		
Water(g)	89.85	
Energy(kcal)	38	
Protein(g)	1.49	
Total lipid (fat) (g)	0.53	
Carbohydrate, by difference(g)	6.86	
Fiber, total dietary(g)		3.8
Sugars, total(g)	1.16	
Minerals		
Calcium, Ca(mg)	15	
Iron, Fe(mg)	3.47	
Magnesium, Mg(mg)	13	
Phosphorus, P(mg)	57	
Potassium, K(mg)	506	
Sodium, Na(mg)	9	
Zinc, Zn(mg)	0.71	
Vitamins		
Thiamin(mg)	0.015	
Riboflavin (mg)	0.215	
Niacin(mg)	4.085	

Vitamin B-6(mg)	*0.044*
Folate, DFE a(Mcg_DFE)	*2*
Vitamin D (D2 + D3) μg	*5.3*
Vitamin D IU	*212*
Ergosterol = 61 mg/100 g.	

Health Benefits of Mushrooms

Mushrooms are a rich source of dietary fiber, and crude proteins. They are rich sources of several types of vitamins. Mushrooms are rich in Vitamin B, C and D. Mushrooms are considered to be rich in the anti-pellagra vitamin, niacin. They are also rich in thiamine (B1), riboflavin (B2), niacin, biotin, and ascorbic acid. Some mushroom varieties are rich in vitamin K and vitamin E also. Mushrooms are rich in minerals particularly calcium, phosphorus, potassium and iron but they are low in sodium. A major percent of the total iron present in mushrooms is in easily available form. Major health benefits of mushrooms are given below:

1. Mushrooms are rich sources of easily digestible crude fiber and hence known as the 'delight of the diabetics'

2. Often grouped with vegetables, mushrooms are rich in nutrients that are commonly found in meat and eggs; hence considered as the 'meat of the vegetable world'

3. The only plant food that contain Vitamin D

4. Low in calories, starch-free, fat-free or very less fat content, and cholesterol-free

5. A good source of B vitamins, including riboflavin, niacin, and Pantothenic acid; Pantothenic acid improves the production of hormones while riboflavin maintains healthy RBCs (red blood cells); niacin promotes healthy skin

6. Among the richest sources of Selenium, an antioxidant that acts as anti-aging and anti-cancerogenic mineral; it also improves the immune system and fertility in men

7. Rich in Ergothioneine, a naturally occurring antioxidant that helps protect the body's cells

8. Rich in Copper that helps make RBCs; it also keeps bones and nerves healthy

9. Rich in Beta-glucans which has immunity-stimulating effects; it also helps resist certain types of allergies

Mushroom Growing Requirements

Choosing a Substrate

Choosing a substrate depends on the type of the mushrooms to be grown. The substrate should be clean and composed of lignin and cellulose and it should be able to hold moisture. Hardwood logs; Hardwood chips and sawdust; seed hulls; leaf litters; paper; cotton wastes; rice straw; wheat straw; and other straws can be used as a substrate for growing most types of mushrooms.

Choosing a Container

Mushrooms can be grown in open straw bales, plastic bags, plastic tubes, and trays of various sizes. As a rule the container should be able to enclose the substrate during the spawn run and should be able to avoid excessive self-heating. It should also allow the maximum production in the space used.

Purchasing Spawns

Purchase the best quality fresh spawns from reputed suppliers. Always read the label on the spawn bottles before purchasing them. One bottle of grain or sawdust spawn in a 500-ml dextrose bottle is sufficient to inoculate 40 to 50 substrate bags (each bag containing approx.

200 kg straw). As a rule, weight of the spawn should be in the range of 1 to 5% of the dry weight of the substrate.

Facilities Required for Mushroom Production

Most important facilities required for mushroom production are,

1. Substrate preparation area

2. Compost turners

3. Pasteurization area and inoculation area (spawning facility)

4. Mushroom growing area

Substrate Preparation Area

Substrate preparation area should be located away from the mushroom growing area. It is better this area has a concrete floor so that substrate can be chopped and mixed well. The substrate preparation area should have,

1. A hammer mill to chop the substrate materials into small pieces

2. A stock of substrate materials

Compost Turners

Now the chopped substrate must be composted using compost turners before it is ready to be used for mushroom growing. The composting process requires that the substrate is kept wet for several weeks.

Pasteurization Area

This area is meant for pasteurization process of composted substrate. It is said that anyone can grow mushrooms if properly prepared substrate is used.

The most important facility required for mushroom growing is a window-less, air-tight building. It can be a metal building covered by polyethylene or a concrete building or a bamboo house insulated with straw and polyethylene sheet. A mushroom growing area should have following requirements:

1.   A temperature of 15 to 20°C (59 to 68°F) and relative humidity of 80 to 95%

2.   Good ventilation and Light: mushrooms are formed when carbon di oxide level is low and hence proper ventilation is required to remove carbon dioxide formed inside the building. Centrifugal blowers may be used to supply air circulation. For light requirements, fluorescent lights are generally used

3.   Proper Hygiene and Sanitation: Air inlets and exits should have a filter to let only filtered air circulated within the building. This is particularly helpful to keep pests and disease-causing pathogens away

4.   Shelves, made from bamboo or wood: These shelves are strong enough to hold substrate bags

Steps Involved in Mushroom Growing

There are **SIX** major steps are involved in the growing of mushrooms. These are,

1.   Compost preparation

2.   Pasteurization

3.   Spawning

4.   Casing

5.   Pinning or Fruiting

6.   Harvesting

Step 1: Compost Preparation

Substrate materials should be chopped into small pieces and wet before composting it. The composted substrate must be pasteurized before being used for spawning. Compost preparation may be done outdoors or indoors provided that proper aeration is present. Major facilities required for successful compost preparation are,

1. A concrete floor to prepare the substrate materials

2. A compost turner to aerate and water the substrate materials

3. A tractor-loader to move the substrate materials to the compost turner

4. Pitchforks to turn the compost piles in the compost turner

The substrate materials are chopped into fine pieces and mixed well with nitrogen supplements and gypsum before wetting them and stacked in rectangular piles. Nitrogen supplements such as chicken manure, soybean meals or peanuts are added to increase the nitrogen content of the compost and gypsum is added to minimize the greasiness compost. Gypsum is added @ 40 lbs. per ton of dry ingredients.

Compost piles are made with tight sides and a loose center for facilitating proper aeration. A recommended size of a compost pile is 5 to 6 feet wide, 5 to 6 feet high, and length as per the requirement. Care should be taken that the compost piles should not be stacked very compact, because air cannot move freely through densely packed compost piles and as a result an anaerobic environment will develop which is detrimental for successful compost preparation process.

Well-prepared substrate materials are then put through a compost turner and water is sprayed at an interval of two days as substrate materials move through the compost turner. As a rule, initial watering is done after there is sufficient heat build-up (145° to 170°F) within the compost piles and watering is done up to the point of leaching while compost pile is turned slowly using pitchforks. After this initial watering, little water is required during the remaining period of composting process. Watering and turning compost piles are done simultaneously. Turning compost piles at regular intervals ensures proper mixing of all the ingredients; proper aeration and relocates the substrate materials from a cooler to a warmer area within the compost pile.

While preparing compost, care must be taken to provide adequate moisture, oxygen, nitrogen, and carbohydrates or else the composting process will fail. Therefore it is essential that water, gypsum and nitrogen supplements are added at regular intervals and the compost pile is adequately aerated as it moves through the compost turner.

Aerobic fermentation of compost begins after a few days and normally a composting cycle lasts from 7 to 14 days depending on the nature of the substrate materials used. Since ammonia and carbon dioxide are produced during composting, a strong smell is associated with composting process. Due to the desirable chemical changes that are associated with the composting process, there are high levels of biological and chemical activity within the compost. This compost is rich in food materials necessary for mushroom growing and it has sufficient heat build-up also. The compost temperatures may reach up to 170° to 180°F during the second and third turnings if desirable composting conditions are maintained.

Towards the end of the composting process, one last turning of the compost piles is done and at this time, water is applied generously up to the point of leaching. Watering is an important process during composting as there is a strong link between watering and the biological activity of the compost. Nutritional value of the compost is enhanced when there is sufficient water to build up a favorable temperature within the piles for the enhanced growth and activity of the beneficial microorganisms.

Characteristics of a well-prepared compost are, a) it has a chocolate brown color; b) it is soft c) it has a moisture content ranging from 68 to 74 percent; and d) it has a strong odor of ammonia

Step 2: Pasteurization Process

Next step required for successful mushroom growing is, pasteurization. Pasteurization is necessary because of two major reasons. These are,

1. Pasteurization is necessary to kill any insects, pests, nematodes and fungal and bacterial infections that may be present in the compost

2. Pasteurization is necessary to remove the excess ammonia present in the compost. Mushroom spawns does not grow well if ammonia concentration in the compost is higher than 0.07 percent

Well-prepared compost is packed into wooden trays before moving them into a specially designed pasteurization chamber. Up to eight wooden trays can be stacked together as a single unit before placing them in the room where an optimum environment is constantly maintained. Alternatively, the bulk system may also be used for pasteurizing compost. Here well-prepared compost is packed in a cement-block bin with a perforated floor and there is no cover on top of the compost bin. While packing compost, care should be taken to allow gas exchange within the compost so that ammonia is replaced by circulating air.

There are two types of pasteurization process: high temperature pasteurization and low temperature pasteurization. High temperature pasteurization involves the raising of the compost temperature up to 145°F for 6 hours. This is done by injecting steam into the pasteurization room where the compost has been placed. After pasteurization process, the temperature is brought down to 140°F by flushing the pasteurization room with fresh cool air. Thereafter, the compost is allowed to cool gradually at a rate of approximately 2° to 3°F each day until all the ammonia is dissipated. During low temperature pasteurization the compost temperature is raised up to 126°F for 6 hours by injecting steam into the pasteurization room, after which the air temperature is brought down to 125° to 130°F range by flushing the pasteurization room with fresh cool air. In the following 4 to 5 days, the compost temperature may be lowered at a rate of approximately 2° to 3°F each day until all the ammonia is dissipated. The pasteurization process can be completed within 10 to 14 days.

At the end of the pasteurization process, the compost temperature should be maintained in the range of 75° to 80°F. The compost should have 2.0 to 2.4 percent nitrogen content and 68 and 72 percent moisture content. Make sure that pasteurization process resulted in as homogenous a material as possible both in terms of nutrient content and temperature. A successful pasteurization process results in 5 to 7 lbs. of dry compost per square foot of bin or tray surface.

In some mushroom growing systems, sterilization is recommended. However Pasteurization is highly desirable over sterilization for successful mushroom growing. Sterilization kills all microorganisms including the beneficial ones present in the substrate while pasteurization does not kill all organisms. Since pasteurization does not kill all the organisms, remaining beneficial organisms present in the substrate indirectly accelerates mushroom formation from the spawns (spores). So the purpose of pasteurization is not to get rid of all organisms, but to get rid of harmful ones while helping to multiply beneficial ones that discourage diseases, consume hemicelluloses, provide nitrogen, and become food for the mushrooms. Pasteurization is very cost-effective also as sterilization process involves expensive high pressure equipments.

Pasteurization is the most critical step in growing mushrooms. The grower must pay close attention to the time and temperature. For large-scale production, automated machines capable of pasteurizing, cooling, spawning and filling the growing containers are available for purchase. Even though these machines are very costly, they provide more protection against diseases and pests, and also save a great amount of human labor.

Step 3: Spawning

The process of inoculating pasteurized compost with the best quality spawns is called inoculation or spawning. It is a good idea to have an inoculation room to prepare spawns before inoculating the substrate bags. If substrate materials are packed in polythene bags, the best method of spawning is either by using grain spawns or by sawdust spawns. If grain spawns are used, spawn bottles are gently shaken to separate mushroom seeds (spores) from mycelia. Thereafter, the bottle is opened carefully and about 2 teaspoon full of spawns is poured into the substrate bag. Soon after inoculation, both the spawn bottles and the substrate bags are covered. To ensure even distribution of spawns throughout the inoculated bag, they are slightly tilted and if sawdust spawns are used, the spawn bottles are broken with an aseptic needle and then a piece of the spawn is transferred carefully to the substrate bags.

Bulk Spawning

Bulk spawning process is practiced in industrial production of mushrooms. Pasteurized mushroom compost is mixed with spawns by hands or by using a spawning machine. Considering the bulk quantities of materials involved, care should be taken to carry out the spawning process in an environment with an excellent sanitation facility. All personnel involved in the process are required to be clean; and wear gloves, face mask and head cap. Inoculation room should have good ventilation but air must be filtered, preferably with a High Efficiency Particulate Air (HEPA) filter. After mixing the mushroom compost with the spawns, they are placed in the growing containers and then the containers are covered. Spawn rate required to inoculate a batch of mushroom compost is often expressed on the basis of spawn weight versus compost weight; a 2 percent spawning rate is considered to be optimum.

Spawn Run

Soon after the inoculation process, the spawned substrate bags are transported into dark mushroom production chambers or mushroom houses. Soon the fungal spores begin to grow and multiply, taking food from substrate materials. The period between the beginning of fungal mycelia growth and the covering of the entire substrate with white-colored mycelium is known as spawn run period and the process is called spawn run. Normally in 15 to 25 days, the entire substrate appears white.

During spawn run period, tremendous heat is generated due to the enhanced biological activity of the growing fungal mycelia and as a result, the compost temperature increases up to 85°F, depending on the type of the mushrooms. This heat build-up adversely affects the growth of the mushrooms as well as the quality of the crop. Therefore it is necessary to maintain proper temperature and moisture content during the entire spawn run period of mushroom growing.

During spawn run, both oxygen and carbon dioxide are required for mycelia growth. Excess water should not be present in the substrate. For this, make sure that sufficient drainage holes are provided in the growing containers. As mycelia grow in number, more oxygen may be required by them; so large holes may be provided on the top of the containers to allow more aeration. Optimum temperature that is to be maintained during spawn run is between 15 and 20°C. There should be adequate ventilation but light is not required.

Step 4: Casing

Casing is the process of uniform-application of a top-dressing material on the spawn-run compost. Normally, materials having good water-holding capacity such as clay-loam field soils or a mixture of peat moss and limestone or similar materials are used for casing.

26

Rhizomorphs or mushroom bodies form on the surface of this casing and mushroom primordia, or pins form on these rhizomorphs. Pasteurized materials should be used for casing to avoid pest-disease infestations.

During spawning and casing, the compost temperature should be maintained in the range of 75° to 80°F and thereafter, the compost temperature may be lowered about 2°F each day until pinning or fruiting occurs. Water must be applied intermittently and a high relative humidity must be maintained before the mushroom pins form.

To increase yield, urea or orchid fertilizer may be used after dissolving it in water @100 grams in 100 liters water. Using a plastic mist sprayer, the solution is sprayed on the surface immediately before fruiting or pinning.

Step 5: Pinning or Fruiting Process

The substrate bags are kept in the mushroom house for another week or so before they are opened to check the mycelia growth. At the time of opening the bags, the bags are supposed to be full with strong growth of mycelium. Generally, fruiting or formation of primordia starts after 3 to 4 weeks of mycelia growth. It is better to make cross-cuts of about 2.5 cm across the bags to allow the mushrooms to grow out. Temperature inside the mushroom house should be increased to range of 20-28oC in order to promote fruiting. Relative humidity should between 80 - 95%. There should be adequate ventilation inside the room to lower the carbon di oxide concentration which should be 0.08 percent or lower.

It is best to place the substrate bags about 10 cm or more above the floor. In doing so, if ventilation is stopped, carbon dioxide accumulated near the floor will not damage the growing mushrooms. It will also restrict the access for insect-pests. Light should be allowed for a few hours a day by keeping doors open. Or else fluorescent lights can be turned on. Light watering of substrate bags using a mist sprayer is recommended daily for higher yields. While watering bags, care must be taken not to overwater the bags. 3 to 4 days after opening the bags, fruiting or pinning starts. Small pins or mushroom primordia develop as outgrowths on a rhizomorph which gradually grow in size into the button stage, and later enlarge in to full-bodied mushrooms. Fruits or mushroom primordia mature in 2 to 3 days once fruiting process is initiated. Generally only one side of the substrate bag is opened at a time. In rare cases, both the sides are opened at the same time.

Step 6: Harvesting Process

Mushrooms are harvested as soon as the gills are well formed and while the edge of the mushroom is still curled under. That is, mushrooms must be picked before they release spores. When the edge flattens and spores are released, the mushrooms lose weight and the spores thus released may cause asthma and hay fever in workers. Mushrooms are harvested by gently pulling them from the substrate. Normally, harvesting is done from the top end of the bag while the other end is just opened to initiate the fruiting process. When fruiting is complete at the lower end, harvesting is done from that end. After harvesting from both the ends, longitudinal slits are made along the central portions of the bag in order to initiate fruiting process. After harvesting from the central portions, check the substrate bag for any presence of developing mycelia. As long as white-colored mycelia are present, fruiting will happen; if the bag appears colorless, it is time to remove them from the mushroom house. Generally the second flush (about 10 days after the first harvest), will be the largest; and it is desirable to destroy the remaining substrate after the third flush. This is recommended because each day in the harvest gives disease-causing pathogens and pests more time to get established in the production area. Once they get established, it is very difficult to keep them away from next crop.

Since mushrooms are harvested by pulling them from substrate materials, it is likely they have a little substrate attached to the stem. Since mushroom stems are not favored among customers, the best practice to clean the product is by trimming the stem.

Preparation for the Next Crop

After the last flush is harvested, the growing room must be cleaned. The traditional method is to inject steam and raise the temperature of the room to 60° and hold it for 4 hours. After that all materials are removed and disposed of. This procedure is particularly important to avoid any disease-pest infestation.

Pests and Diseases in Mushroom Cultivation

Diseases, such as Bacterial Blotch, Green Mold; Verticillium Dry Bubble and Mildew are generally eliminated during pasteurization process. Proper sanitation of mushroom growing facilities will keep almost all diseases away for a long period. Major pests that are found to affect mushrooms are, Sciarid Flies and Phorid Flies. Insect and pests can be controlled by using mechanical traps, insect-proof nets and other biological control methods.

Yield

Generally 100 kg dry weight of substrate yields 200 Kg mushrooms. That is, yield ranges from about 150-200 % of the dry weight of the substrate depending upon the type of substrate materials used and cultural practices adopted.

Maturity Indices

A ripe mushroom will have a glossy, uniform and a well-rounded cap and tight gills. Maturity is reached when the caps are well- formed and the gill is completely intact.

Quality Indices

Major quality parameters that reconsidered for mushroom marketing are size of the cap, color, visibility of the gills, presence of stipe or stem and cleanliness of the product. An illustration of quality indices for mushrooms is given below:

Quality Index	Description
Size of the mushroom cap	Uniform, well rounded cap
Color of the entire mushroom	1. Cap with a smooth glossy surface and fully intact gills
	2. Discolored mushrooms with spots are of inferior quality
Gills	1. Gills should be intact
	2. Gills should not be open and visible
	3. Mushrooms with visible gills are of inferior quality
Cleanliness	No or minimum growth medium residue

Source: UC Davis, USA

Physiological & Physical Disorders

Common physiological disorders that are found in mushrooms are upward bending of caps; opening or visibility of the gills; bruises and damages due to rough handling; browning discoloration; freezing injury and CO_2 injury. Freezing injury occurs at temperatures of -0.6°C (30.9°F) or lower. Signs CO_2 injury are blackening and pitting.

Storage of Mushrooms

Since mushrooms are highly perishable, i.e. at room temperature, maximum shelf of mushrooms is a day or two, they must be precooled at 3 – 5°C (37 - 41°F) as soon as they are harvested in order to prolong their shelf life. Cost effective method of cooling is to place freshly-harvested mushrooms in a cool vacuum chamber. Another cooling method is by mechanical refrigeration. Cooling equipments may also be used for faster cooling. After precooling, mushrooms may be stored for a considerable period at optimum temperature and humidity. Optimum temperature for storage is considered to be 0° - 1.5°C (32° - 35°F). Storage life is typically 5-7 days at 1.5°C (35°F) and 2 days at 4.5°C (40°F). Optimum Relative Humidity for storage is 95-98%. High relative humidity is essential to prevent desiccation and loss of glossiness of mushrooms.

Packaging

Generally mushrooms for shipping are packed in bulk quantities and shipped in cartons with a perforated overwrap to maintain high humidity. Center-loading during shipment improves air circulation necessary for mushrooms. Mushrooms for retail markets are packed in plastic trays or paper trays in small quantities ranging from 200 gm to 500 gm and overwrapped with a thin plastic film to conserve moisture inside.

Marketing of Mushrooms

Mushrooms are marketed as fresh, frozen, canned, marinated, freeze-dried and tunnel-dried mushrooms.

Freeze-Dried Mushrooms

Freezing the mushrooms, then placing them in a vacuum where they remain frozen until all water is removed. That process is called freeze-drying and is very expensive, both for energy and for the required equipment.

Tunnel-Dried Mushrooms

A tunnel drier consists of a blower to circulate air, a heater to increase the temperature of the air to approximately 40 to 50°C (104 to 122°F), and a chamber to put the food to be dried. Tunnel drying will give a high quality product.

Bibliography

Arora, D. (1986). *Mushrooms Demystified* (p. 1056 pages). Ten Speed Press.

Kuo, M. (2007). *100 Edible Mushrooms* (p. 344 pages). University of Michigan Press.

Schwab, A. (2007). *Mushrooming without Fear: The Beginner's Guide to Collecting Safe and Delicious Mushrooms* (p. 128 pages). Skyhorse Publishing.

Shu-ting Chang, P. G. (2004). *Mushrooms: Cultivation, Nutritional Value, Medicinal Effect, and Environmental Impact* (p. 451 pages). CRC Press.

Thomas Laessoe, A. D. (1996). *The Mushroom Book How to Identify, Gather and Cook Wild Mushrooms and Other Fungi* (p. 256 pages). DK ADULT.

www.ingramcontent.com/pod-product-compliance
Lightning Source LLC
Chambersburg PA
CBHW071558170526
45166CB00004B/1710